LET'S PLAY SCIENCE

LET'S PLAY SCIENCE

WRITTEN AND ILLUSTRATED BY MARY STETTEN

HARPER COLOPHON BOOKS
Harper & Row, Publishers
New York, Hagerstown, San Francisco, London

LET'S PLAY SCIENCE. Copyright © 1979 by Mary Stetten. All rights reserved. Printed in the United States of America. No part of this book may be used or reproduced in any manner whatsoever without written permission except in the case of brief quotations embodied in critical articles and reviews. For information address Harper & Row, Publishers, Inc., 10 East 53rd Street, New York, N.Y. 10022. Published simultaneously in Canada by Fitzhenry & Whiteside Limited, Toronto.

FIRST EDITION

ISBN: 0-06-090711-8

Library of Congress Catalog Number: 78-24698

80 81 82 83 84 10 9 8 7 6 5 4 3 2 1

THIS BOOK IS LOVINGLY DEDICATED TO THE TEACHERS AND CHILDREN OF INTERNATIONAL PLAY GROUP

ALSO... TO MOM, DAD AND MICHAEL ♥

About the Author/Illustrator

MARY STETTEN received early scientific schooling as a child at the Woods Hole Children's School of Science on Cape Cod. She earned her bachelor's degree at Oberlin College and a Master of Science in early childhood education at the Bank Street College of Education.

Ms. Carson runs a science program at International Play Group in New York City and teaches a workshop on science for preschoolers at the Bank Street College.

CONTENTS

GROWING THINGS
- 2 WATCH A SEED START TO GROW
- 4 GROW (AND EAT) BEANSPROUTS
- 6 START A SEED COLLECTION

FOUR THINGS THAT GROW
- 8 CARROTS
- 9 POTATOES
- 10 ONIONS
- 11 INDIAN CORN

TWO TYPES OF LEAF COLLECTION
- 12 PRESS LEAVES
- 13 PRINT LEAVES

USE YOUR BODY

TOUCH
- 16 PLAY A TOUCH GAME
- 18 MYSTERY TOUCH BOX

TASTE
- 20 PLAY A TASTE GAME

SMELL
- 22 PLAY A SMELL GAME

HEAR
- 24 PLAY A MUSIC GAME
- 26 MAKE A RATTLE

28	MAKE A KAZOO
30	MAKE A STRINGED THING
32	MAKE A HOSE TELEPHONE

LIGHT, SHADOWS AND AIR

LIGHT
| 36 | MAKE A RAINBOW |

SHADOWS
38	PLAY A SHADOW GAME
40	MAKE A SHADOW THEATRE
42	MAKE A SHADOW PICTURE

AIR
44	MAKE A JET
46	MAKE A PINWHEEL
48	MAKE A PAPER AIRPLANE
50	MAKE A HELICOPTER
52	MAKE A PARACHUTE
54	STRAWS AND AIR

BALANCES, MAGNETS, PULLEYS, MACHINES, WATER AND MORE

TWO BALANCES YOU CAN MAKE
58	MAKE A COAT HANGER BALANCE
60	MAKE A BALL BALANCE
62	PULLEYS
64	USE A GRINDER TO MAKE PEANUT BUTTER

WATCH A SEED START TO GROW

Try to sprout several types of seeds. Go on a collecting expedition outside and bring home as many types of seeds as you can find. Try sprouting these seeds. Don't be disappointed if not all of them sprout. Some types of seeds require special conditions, such as a long period of cold weather, before they can grow.

After a seed has started to grow you can transplant it to a pot full of soil. Keep the plant in a sunny place and water it every few days.

You will need:

Dry beans such as lima or kidney beans. Seeds from fruits and vegetables or collected outdoors.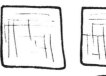

Paper towels

A clear glass or plastic cup

Fill the cup loosely with several sheets of crumpled paper towels.

Put several seeds into the cup, between the paper and the glass, so that you can see them through the glass.

Drip water onto the paper towels until they are moist.

Keep the towels damp by watering every day. In a few days the seeds should start to grow.

GROW and eat BEANSPROUTS

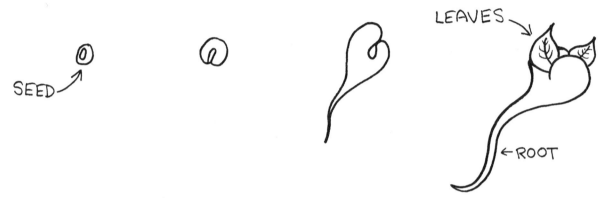

You will need:

Dry beans or seeds - including: chickpeas, flax seeds, lentils, mung beans, soy beans, alfalfa seeds (available at grocery and health food stores).

An empty glass jar
or
A Mason jar with a lid that has a removable center.

Cheesecloth cut into a square large enough to fit over the top of the jar.

A rubber band (if using a plain jar).

Use enough seeds to form a thin layer on the bottom of the jar. Rinse the seeds and soak overnight in warm water.

Fasten cheesecloth over the top of the jar. (If you use a Mason jar, remove the center section of the lid and use the outer ring to hold the cloth in place.) Pour off the water.

Rinse the seeds in cool water and lay the jar in a dark place. Rinse the seeds two or three times a day. The sprouts will be ready to eat in three or four days. To green up the sprouts, place the jar in the sunlight for a day.

What did the seed need before it could start to grow? How did the seed change as it grew?

The beansprout is a baby plant. Can you find its root? Are there any leaves?

Many people cook with beansprouts, but they are also good to eat plain or mixed with a salad.
    To make a beansprout salad:
      Wash and tear up some lettuce.
      Wash and cut up some tomatoes, carrots, onions and whatever else you like in salad.
      Mix in the beansprouts and toss with salad dressing.

START A SEED COLLECTION

SEEDS IN THINGS YOU EAT

SEEDS YOU CAN FIND OUTDOORS

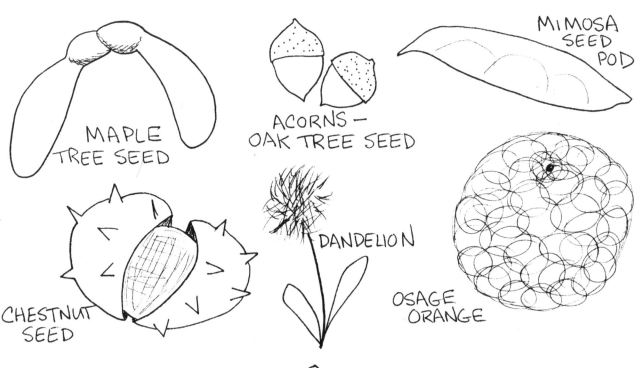

You will need:

Empty egg cartons (for sorting the small seeds) and a box (for the larger seeds).

A hammer, scissors and knife for opening seeds.

A paper bag for collecting seeds.

Go on a collecting expedition in your neighborhood. Bring along a paper bag and collect as many types of seed as you can find. Collect seeds from the fruits and vegetables that you eat at home.

Open some of the seeds that you collect. Hard seeds, such as nuts, can be opened with a hammer. Other seeds can be opened with scissors, a knife or your fingers.

Seeds come in many shapes, sizes and textures. Some seeds are soft and fluffy and can travel great distances on the wind. Nuts have hard shells that protect the seed inside. Can you guess why each seed is the way it is?

Try to sprout the seeds you have collected. (See "Watch a Seed Start to Grow.")

Use the seeds in art projects by pasting them on cardboard. Make up a seed sorting game.

Find out the names of the seeds you found by looking for their pictures in a book about trees.

GROWING THINGS FROM THE KITCHEN

CARROTS

You will need:

A fresh carrot with a green top

A dish, Pebbles, A knife

Cut off the top one or two inches of the carrot.

Put a thin layer of pebbles into the dish, place the carrot on top and add enough water just to cover the base of the carrot. Keep in indirect sunlight and add water as needed.

POTATOES

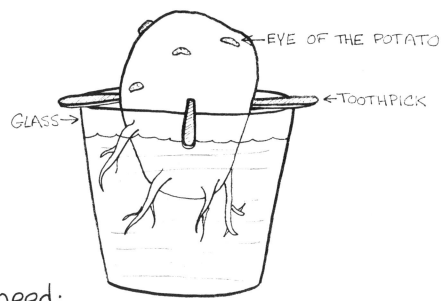

You will need:

A potato (white or sweet) with a lot of eyes

A glass, Toothpicks

Poke toothpicks into the potato and place it in a glass of water with the <u>narrow</u> end down.

Keep white potatoes in a cool, dark place. Keep sweet potatoes in the sunshine Add water as needed.

Transplant to soil after one or two months.

ONION

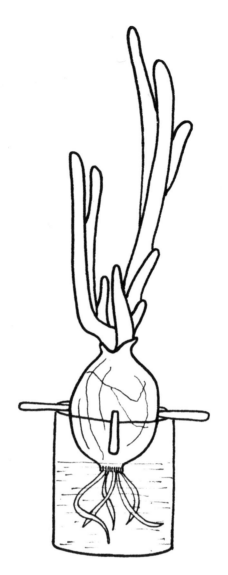

You will need:

    An onion. Try to pick one that has some roots growing at one end.

    Four toothpicks, A glass

Poke the toothpicks into the onion and place it in a glass with the root end down. Add water just to cover the roots.

INDIAN CORN

(A project for just after Thanksgiving)

You will need:

An ear of Indian corn

A long shallow dish - such as an ice cube tray.

Lay the corn in the dish and add about one inch of water. Change the water every other day.

2 TYPES OF LEAF COLLECTION

PRESS LEAVES

Collect many types of leaves. For beautiful colors collect in the Fall.

Lay the leaves between sheets of paper. To press a lot of leaves, alternate leaves and paper.

Lay a heavy stack of books on the paper and press for one week.

Paste leaves on heavy paper.

PRINT LEAVES

Lay leaves on newspaper with the vein side up. Paint the upper side of the leaf with printers ink or tempera paint that has been thinned with water.

Transfer the leaves carefully to a clean sheet of paper, keeping the painted side up. Lay another sheet of paper on top of the leaves and press with your hand. Lift up the top sheet of paper and carefully peel off the leaves.

USE YOUR BODY

THE RABBIT

- EARS
- EYES
- NOSE
- MOUTH
- HAND
- TUMMY
- FOOT

TOUCH

PLAY A TOUCH GAME

You will need:

Cardboard , A blindfold

Scraps of sandpaper, sponge, cloth, etc.

Scissors    Glue

Cut the cardboard into rectangles about the size of playing cards. Glue a different textured material to one side of each card.

Put on a blindfold and have a friend select a card for you to feel. How does the material feel? Soft? Bumpy? Rough? Smooth? Return the card to its place.

Remove the blindfold and try to guess which card you felt.

Your skin is covered with tiny nerve endings that send messages to your brain. Your brain can then tell you how a thing feels.

Try playing the Touch Game with different parts of your body. Can you feel just as well with your nose? Your toes? Your elbow?

How many types of feels can you name? Think of some things that feel hot. Cold. Prickly. Soft.

Cut an opening in the box large enough for a hand to fit through. Decorate the box with felt-tipped pens, paints or collage.

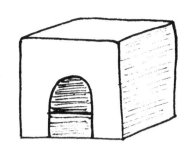

Put a different object in each sock and tie the sock closed.

Put all the socks into the box.

To play: Take a sock from the box, feel it. What do you think is inside? Open the sock and find out if you guessed correctly

"I WAS RIGHT"

How did you guess what was in the sock? Try to describe how it felt. Was it soft? Hard? Smooth? Bumpy? What was its shape?

TASTE
PLAY A TASTE GAME

Look at your tongue in the mirror. If you look very carefully you will see tiny bumps on the top side. These bumps are called taste buds, and they are what you use for tasting.

What are some of your favorite foods? Can you describe how they taste?

You will need:
- Many different types of food, cut into bite size pieces

  GRAPE CHEESE PEANUT SALAMI APPLE COOKIE CEREAL ORANGE

- Cups - one for each type of food

- A blindfold

Put on the blindfold and have a friend put one type of food in your mouth.

How does it taste? How does it feel? Is it...

sweet?   sour?   crunchy?

chewy?   spicy?

Can you guess what it is?

SMELL
PLAY A SMELL GAME

Your nose can smell many things. Have you ever smelled a fire burning? Or gasoline at a gas station? Does your Mommy wear perfume? Sometimes you can even tell what is cooking for dinner by its smell.

Your nose is good at smelling, but some animals have even better noses. When you meet a dog, he will want to sniff you to get to know your smell. There is a type of dog called a bloodhound that is famous for being able to find people by following the scent they leave behind.

Don't try to play the Smell Game if you have a bad cold. When your nose is stuffy it is often very hard to tell one smell from another.

You will need:

Empty plastic containers with lids

Foods with distinct smells

CHOCOLATE CHIPS   LEMON   CHEESE   ONION   APPLE   VANILLA   PEANUT BUTTER

Put a small amount of food in each container and put on the lids.

Allow them to stand for a while at room temperature so that the smells can grow strong.

Open one container a crack and sniff. Do not peek at the food inside. Can you guess what the food is by its smell? Look inside and find out if you were right.

HEAR
PLAY A MUSIC GAME

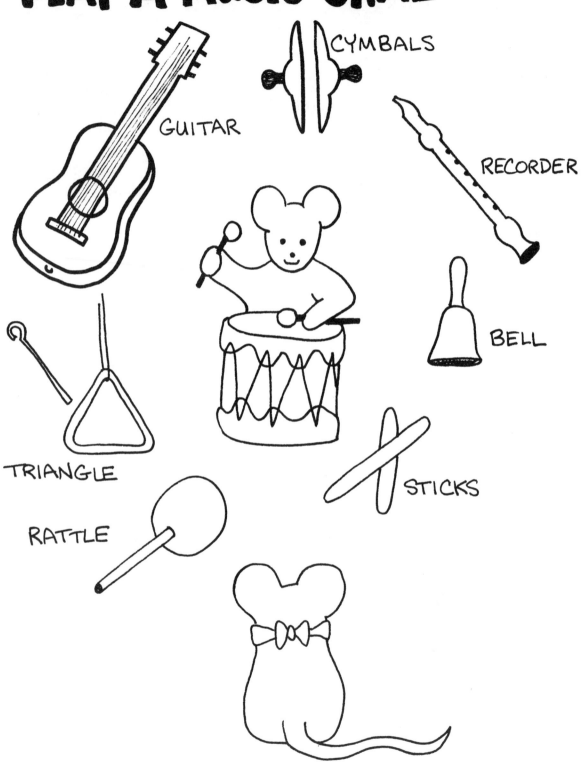

You will need:

    Musical instruments - such as a bell, a rattle, a guitar or a drum.

    A blindfold

Put on a blindfold. Have a friend play one of the instruments. How did it sound? Can you guess which instrument you heard?

What words can you use to describe how the instrument sounded? Did it bang or crash or ring or tap or tinkle?

How many ways can you find to play the instrument? Can you play it loud? softly? fast? slow?

When you listen you are using your ears. Can you hear well if you have your fingers in your ears?

Sit very still for a minute or two. How many noises can you hear? Did a car honk its horn? Did a dog bark? Are people talking to each other in the next room?

Try playing the Music Game with instruments you have made yourself. Make up your own instruments or follow the instructions on the next few pages.

MAKE A RATTLE

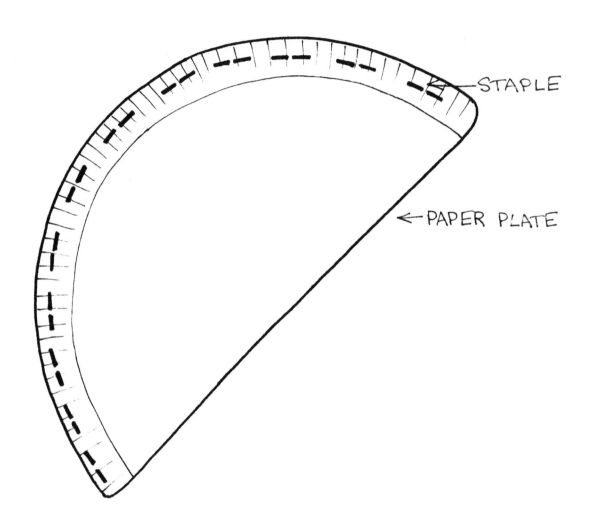

STAPLE

← PAPER PLATE

You will need:

Dry beans or pebbles

A paper plate

A stapler

Crayons or felt-tipped pens

Fold the paper plate in half

Drop in some beans or pebbles

Staple the plate together near the edge and decorate with crayons or felt-tipped pens.

PLAY A RATTLE GAME

Make several pairs of rattles, filling each pair with the same material.

  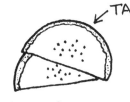 

FILLED WITH BEANS FILLED WITH PENNIES FILLED WITH SAND FILLED WITH PAPER CLIPS

On one side of each rattle draw a picture of what is inside.

Mix the rattles together with the picture sides down.

Find the matching pairs by shaking the rattles and comparing the sounds. Check by looking at the pictures.

MAKE A KAZOO

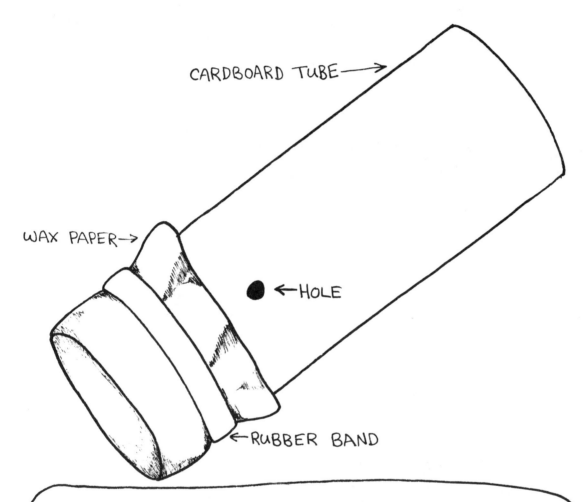

Why does the kazoo make such a funny sound?

Place your finger gently on the wax paper while you play the kazoo. What do you feel? The wax paper tickles your finger because it is moving back and forth very quickly.

All sounds are caused by something vibrating. When you hit a drum the top of the drum vibrates. When you pluck a guitar string the string vibrates.

Make a lot of kazoos so that all your friends can play together. Form a kazoo orchestra and play all your favorite songs.

You will need:

- A cardboard tube from a roll of toilet paper or paper towels.

- A piece of wax paper large enough to cover one end of the tube

- A rubber band

- A pencil

Cover one end of the tube with wax paper and secure it with a rubber band.

Use a pencil to poke a hole in the tube near the wax paper.

To play the kazoo: Put the open end to your mouth and sing your favorite song substituting "du-du-du" for the words.

MAKE A STRINGED THING

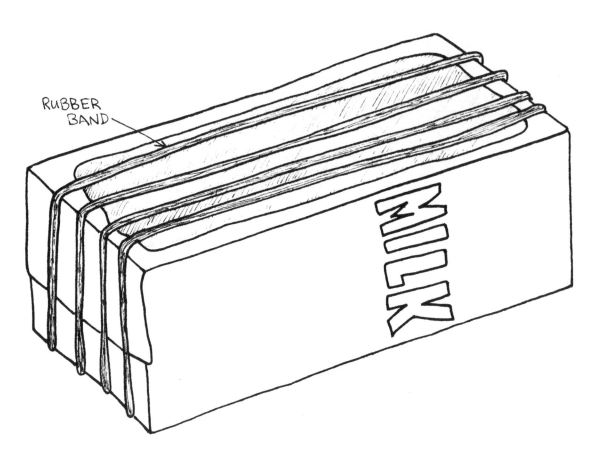

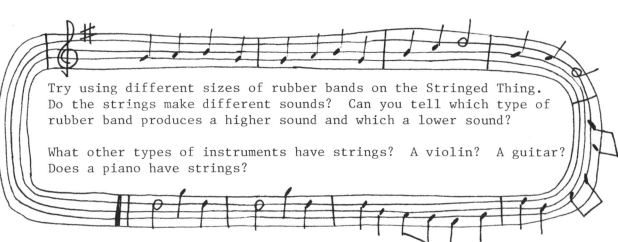

Try using different sizes of rubber bands on the Stringed Thing. Do the strings make different sounds? Can you tell which type of rubber band produces a higher sound and which a lower sound?

What other types of instruments have strings? A violin? A guitar? Does a piano have strings?

You will need:

   An empty milk carton

   Rubber bands

   Scissors

Wash out the milk carton and open up the top. Cut down the sides to form four flaps.

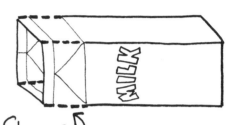

Cut out one side of the milk carton, leaving a small rim on all four sides.

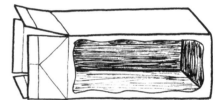

Fold the open end of the carton so that it lies flat.

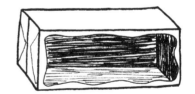

Stretch several rubber bands around the carton.

MAKE A HOSE TELEPHONE

Can you guess how the hose telephone works?

You may think that the hose is empty, but it is not; it is filled with air. When you talk to one end of the telephone the air starts to vibrate, which means that it moves back and forth very quickly. The vibrating air travels along the hose until it comes out into your friend's ear.

This type of phone was used long ago in big houses so that people could talk to each other from different rooms. It is still a common type of telephone on ships.

Try using the phone to talk to a friend in another room. Use the phone with several friends and try to guess whom you are talking to. But remember, never yell through a hose telephone; the loud noise might hurt your friend's ear.

You will need:

A long length of hose (10, 20 or even 30 feet). Buy hosing at a hardware store or use an old garden hose with the ends cut off.

Two plastic containers which are open at one end.

Cut a cross in the base of the container.

Push one end of the hose through the slit and out of the open end of the container. Wrap tape several times around the hose near the end.

Pull the hose back into the container.

Repeat with the other end of the hose.

LIGHT
MAKE A RAINBOW

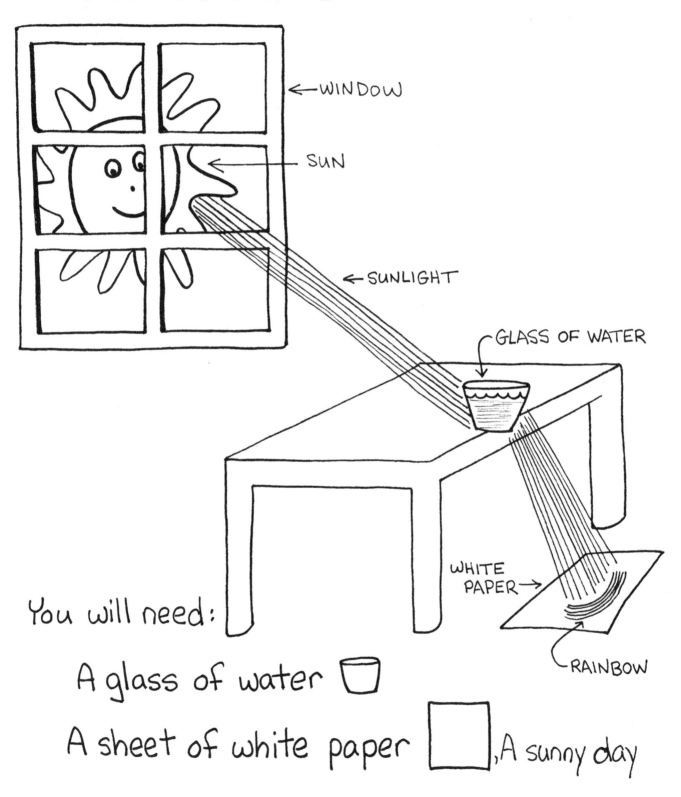

You will need:

A glass of water

A sheet of white paper , A sunny day

Fill the glass almost to the top with water. Place the glass so that it is half on and half off the edge of a table and so that the sun shines directly through the water and onto a sheet of white paper on the floor.

Adjust the paper and the glass until a rainbow forms on the paper.

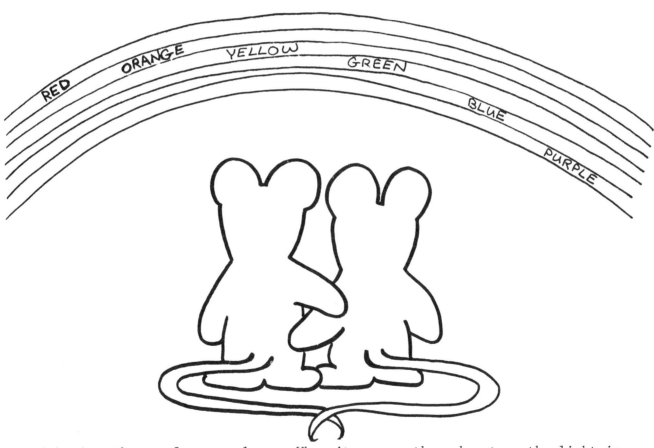

Light is made up of many colors. When it passes through water, the light is broken, and these colors can be seen as a rainbow.

Have you ever seen a rainbow in the sky? Rainbows usually appear when the sun comes out just after a rain. Why is this?

Try to make a picture of a rainbow with crayons or paint. Be sure to put the colors in the right order.

SHADOWS

PLAY A SHADOW GUESSING GAME

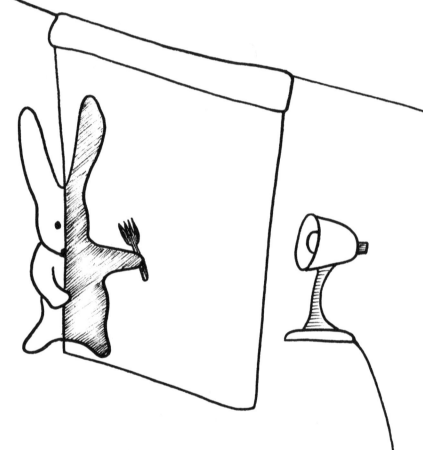

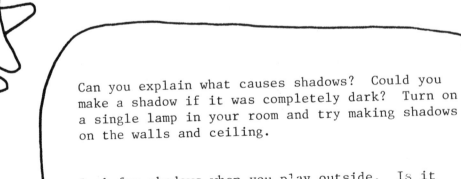

Can you explain what causes shadows? Could you make a shadow if it was completely dark? Turn on a single lamp in your room and try making shadows on the walls and ceiling.

Look for shadows when you play outside. Is it easier to see shadows on a sunny or a cloudy day?

You will need:

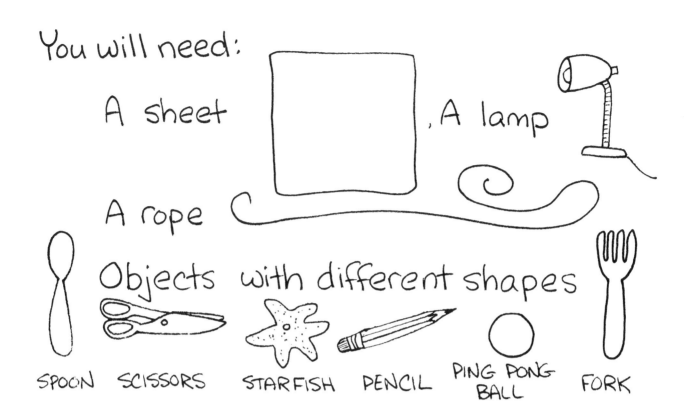

A sheet, A lamp

A rope

Objects with different shapes

SPOON SCISSORS STARFISH PENCIL PING PONG BALL FORK

String the rope across the room and hang the sheet over it. Have your friends sit on one side of the sheet and shine the light on the other side. Darken the room.

Stand between the light and the sheet and hold up one of the objects so that the shadow shows on the other side of the sheet. Have your friends try to guess what the object is by the shape of its shadow.

Move the object toward the light. What happens to the size of the shadow it casts on the sheet.

MAKE A SHADOW THEATRE

You will need:

- A rope
- A sheet
- A lamp
- Heavy paper or cardboard
- Pipe cleaners
- Masking Tape
- Scissors

Cut figures of animals, people, trees, etc. from heavy paper or cardboard.

Tape a bent pipe cleaner to the back of each figure.

Set up the sheet and light as you did for the Shadow Game. Darken the room.

Move the puppets behind the sheet. Make the shadows big or small by moving the puppets back and forth between the sheet and the lamp.

MAKE A SHADOW PICTURE

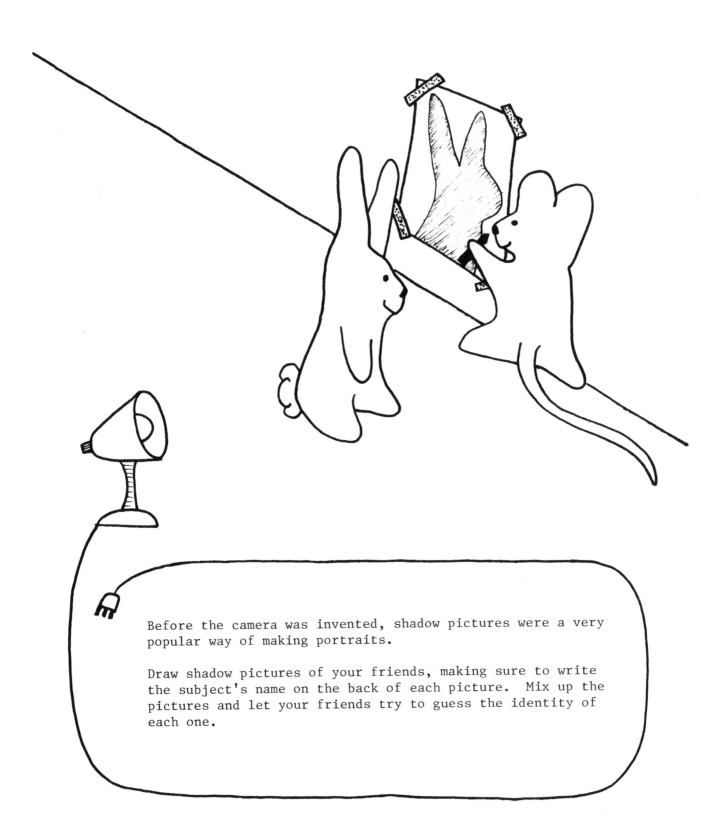

Before the camera was invented, shadow pictures were a very popular way of making portraits.

Draw shadow pictures of your friends, making sure to write the subject's name on the back of each picture. Mix up the pictures and let your friends try to guess the identity of each one.

You will need

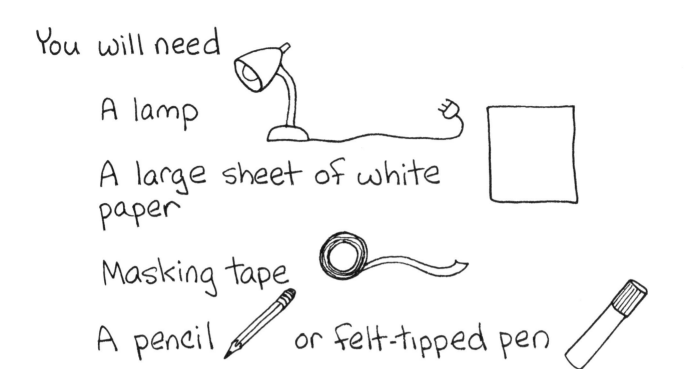

A lamp

A large sheet of white paper

Masking tape

A pencil or felt-tipped pen

Have your friend stand near a wall and shine a light so that the shadow of the face's profile appears on the wall.

Tape a sheet of paper on the wall and adjust the light and your friend so that the shadow fits onto the paper.

Carefully trace the outline of the shadow with a pencil or felt-tipped pen. Your friend must stand <u>very</u> still for you to draw a good portrait.

Remove the paper from the wall and write your friend's name on the back.

AIR
MAKE A JET (WITH A BALLOON)

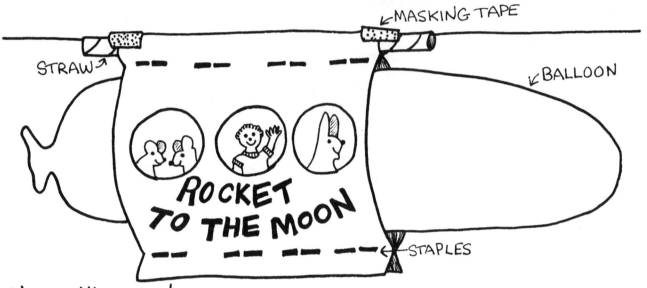

You will need:

A piece of string - long enough to reach across the room

Several long balloons

A drinking straw

A piece of construction paper

Felt-tipped pens or crayons

A stapler    Masking tape

Thread the string through the straw and tie the string so that it stretches tightly across the room.

Fold the construction paper in half and decorate it to look like a rocket ship.

Hang the folded paper over the straw and tape it in place.

Staple the paper together just under the straw and again near the bottom.

Attach a deflated balloon to the inside of the paper with a roll of masking tape so that the opening of the balloon sticks out from the back end of the rocket. Blow up the balloon and release to fly the rocket.

What makes the balloon rocket go?

What will happen if you blow up the balloon and release it without attaching it to the rocket?

What would happen if you used a smaller balloon? A larger balloon? Does the amount of air in the balloon have anything to do with how far the rocket will travel?

10, 9, 8, 7, 6, 5, 4, 3, 2, 1, BLAST OFF!

MAKE A PINWHEEL

You will need:

A square piece of paper (6in. x 6in.)

Scissors, A pin

A stapler

Masking tape

A straw

Fold the paper into a triangle

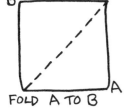
FOLD A TO B

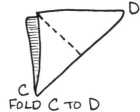
FOLD C TO D

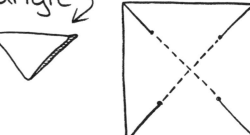

Unfold, and cut along the fold lines half way to the center.

Fold the corners of the paper (X) to the center (Y). Bend but do not fold the paper.
Staple the pinwheel together at the center point.

Stick a pin through the center of the pinwheel (front to back) and into the straw. Cover the point of the pin and the top of the straw with masking tape. Make sure to cover the pin on both sides of the straw. The tape should hold the paper away from the straw.

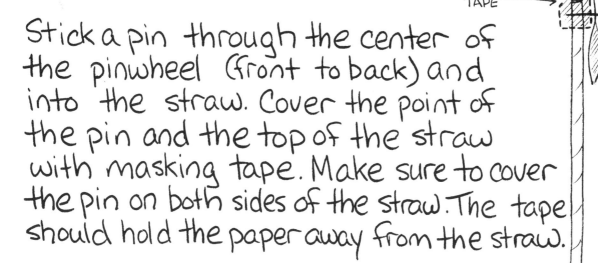

What makes the pinwheel move?

You cannot see air, but when you blow, the air moves and pushes the pinwheel round and round. Moving air can also push things like sailboats, windmills and balloon jets (see MAKE A JET WITH A BALLOON).

Try to make the pinwheel go around without blowing on it. One way is to move the pinwheel through the air by running with it. Try holding the pinwheel in front of a fan or out a car window.

MAKE A PAPER AIRPLANE

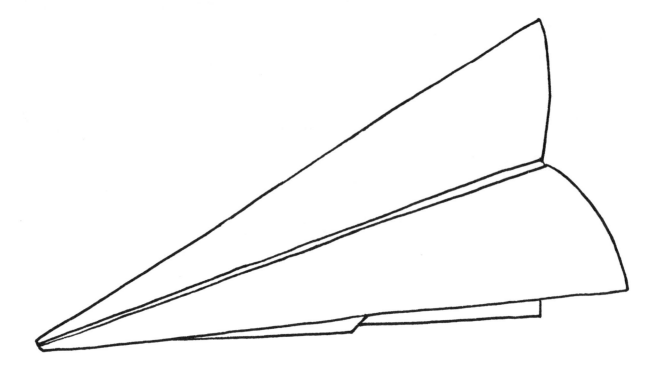

You will need:

A piece of paper

A stapler

Scissors

Try to throw a piece of paper that has not been folded. How does it fly? Can you explain why the folded airplane flies better than the plain sheet of paper?

There are many ways of folding paper into airplanes. Fold and test your own models.

Decorate your planes with felt-tipped pens.

Fold a piece of paper in half lengthwise. Open up the paper and lay it on the table with the crease up.

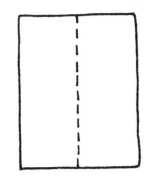

Fold the two top corners till they meet at the center crease.

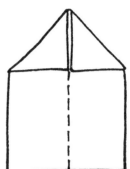

Fold again to form a sharper point.

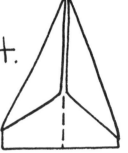

Fold the plane on the original crease. Staple in the center of the plane near the center crease.

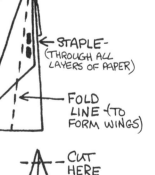

← STAPLE (THROUGH ALL LAYERS OF PAPER)

← FOLD LINE (TO FORM WINGS)

-- CUT HERE

Form the wings by folding them to the center crease on both sides.

Snip off the point of the plane.

MAKE A HELICOPTER

You will need:

A piece of paper

Scissors

Crayons or felt-tipped pens

Crease the paper down the center lengthwise; unfold.

Fold points A+B until they meet on the center crease.

Fold points C+D till they meet on the center crease.

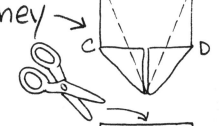

Cut down the center crease to point F.

Fold the two flaps slightly in each direction.

Drop the helicopter from a high place.

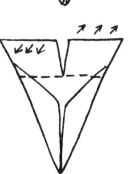

Try dropping the paper at each stage of folding. How does it fall before you start to fold? How does it fall with a point but no flaps? Does the angle of the flaps affect how the finished helicopter falls?

Decorate the helicopter with felt-tipped pens.

MAKE A PARACHUTE

Drop the spool or clothespin without a parachute. How does it fall?
Can you explain how the parachute works?

To make a beautiful parachute, tie die the fabric or decorate it
with felt-tipped pens.

You will need:

- A square piece of cloth (A handkerchief is good)
- String, Scissors
- A clothespin, spool or small doll ← DRAW ON FACES

Cut four pieces of string of equal length.

Tie one piece of string to each corner of the cloth.

If you're using a spool, thread all four strings through the hole and tie them together.

If you're using a clothespin, tie the four strings together and then tie on the clothespin with a separate piece of string.

AIR AND STRAWS
BLOW BUBBLES

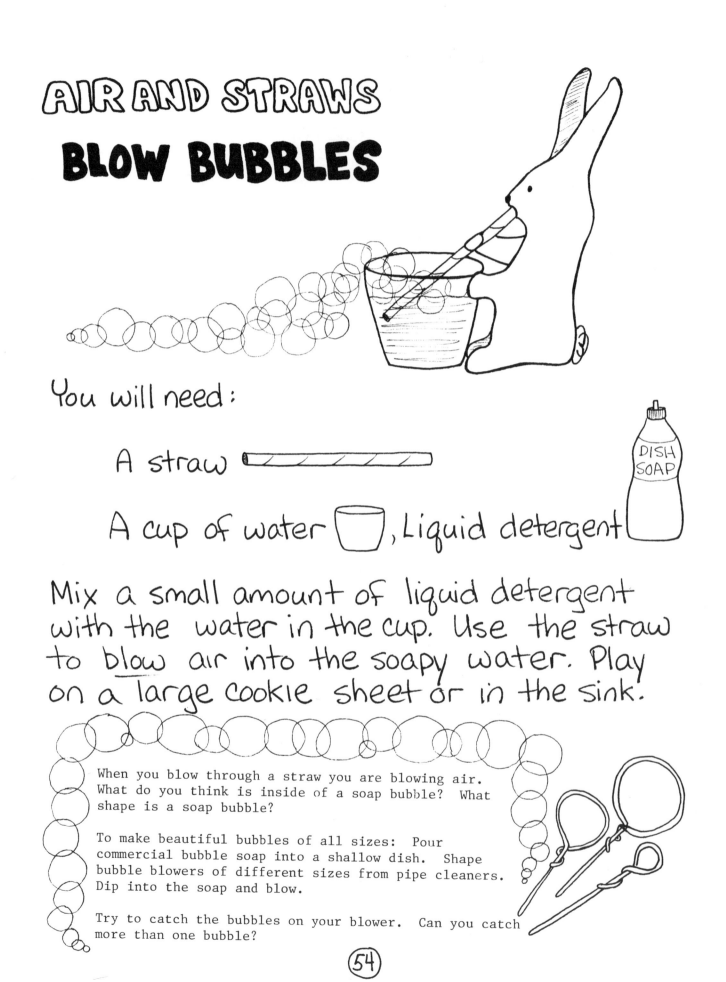

You will need:

A straw

A cup of water, Liquid detergent

Mix a small amount of liquid detergent with the water in the cup. Use the straw to blow air into the soapy water. Play on a large cookie sheet or in the sink.

When you blow through a straw you are blowing air. What do you think is inside of a soap bubble? What shape is a soap bubble?

To make beautiful bubbles of all sizes: Pour commercial bubble soap into a shallow dish. Shape bubble blowers of different sizes from pipe cleaners. Dip into the soap and blow.

Try to catch the bubbles on your blower. Can you catch more than one bubble?

MAKE A BLOW PICTURE

You will need:

Ink or tempera paint thinned with water

A straw, white paper

Put a large drop of paint on the paper. Blow at it through the straw. Continue blowing at wet spots and turning the paper to create a picture. Add several colors.

2 BALANCES YOU CAN MAKE
MAKE A COAT HANGER BALANCE

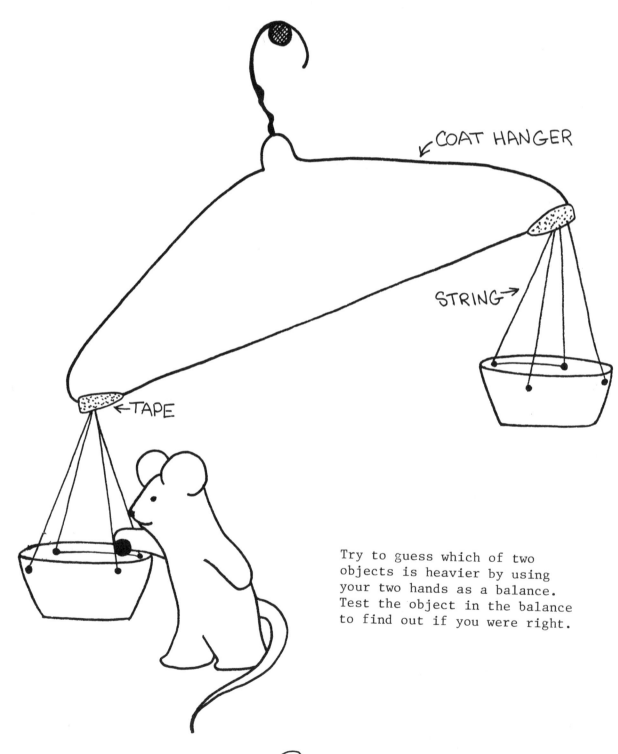

Try to guess which of two objects is heavier by using your two hands as a balance. Test the object in the balance to find out if you were right.

You will need:

A coat hanger, String,

Two identical plastic containers,

A nail, Scissors, Masking tape

Cut 4 lengths of string, each one yard long.
Poke four holes in each container with the nail. The holes should be evenly spaced near the rim.

Thread each string through two of the holes.

← TAPE WRAPPED AROUND THE END OF STRING WILL MAKE IT EASY TO THREAD THROUGH THE HOLES.

Tie the four strings together near the ends.

Tie a container to each end of the hanger and secure to the hanger with masking tape.

← TAPE

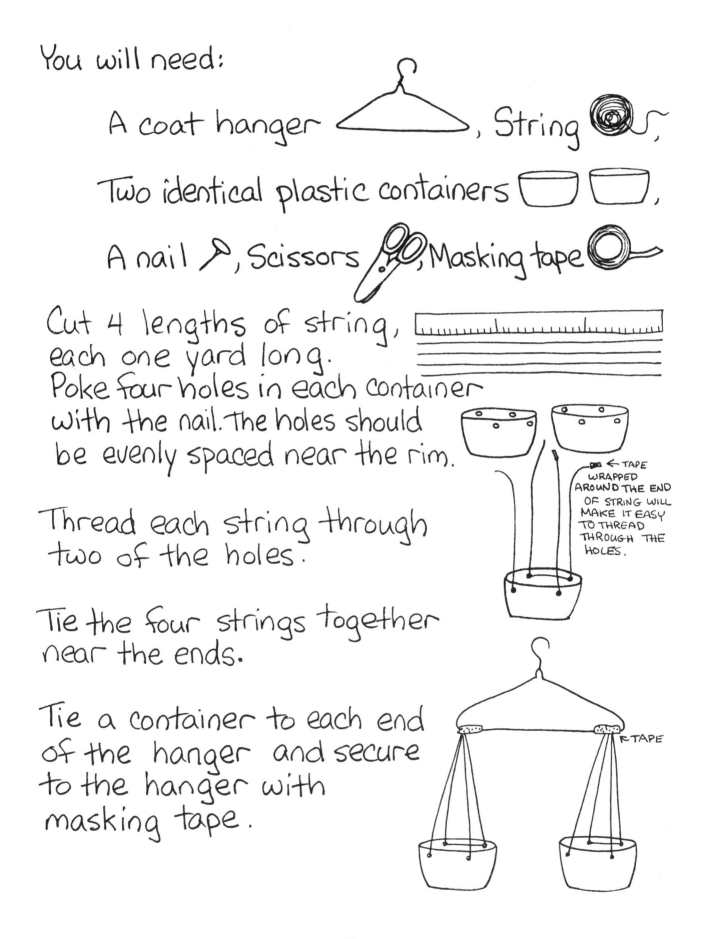

MAKE A BALL BALANCE

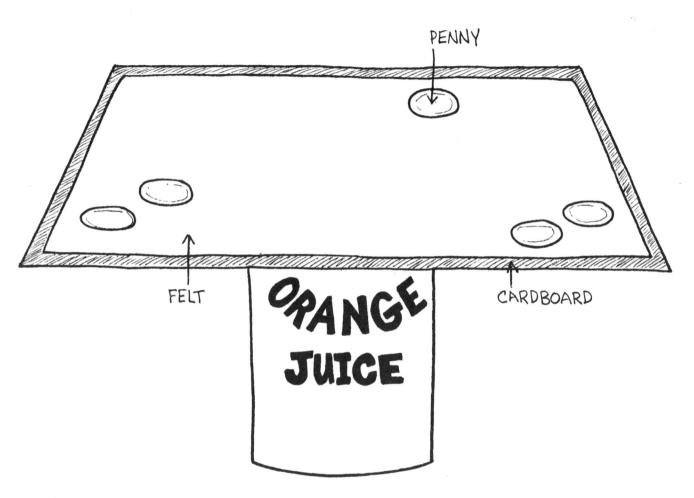

You will need:

An empty can , Newspaper

A rubber ball (small enough to fit into the can)

Heavy cardboard , Scissors

Felt , Glue , Pennies

Fill the can most of the way up with crumpled newspaper and set a ball on the newspaper.

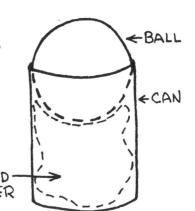

Cut the cardboard into a square and cut two pieces of felt the same size. Glue the felt to both sides of the cardboard.

Balance the board on the ball. Use pennies as weights, placing them on the board so that the board tilts or remains level.

Play a balance game: Take turns placing coins on the balance. The player who makes the board fall is the loser. How many coins can you balance?

⇨ Make balance boards in other shapes.

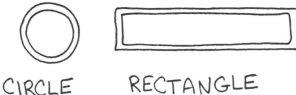

PULLEYS

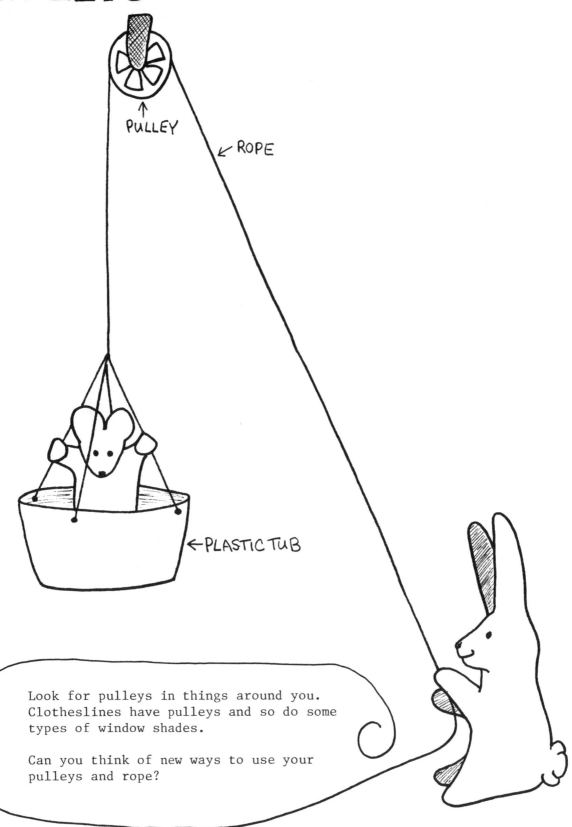

Look for pulleys in things around you. Clotheslines have pulleys and so do some types of window shades.

Can you think of new ways to use your pulleys and rope?

You will need:

Pulleys - clothesline pulleys are available at the hardware store.

A small plastic container

String Scissors  A nail

Use the nail to poke four holes in the container. The holes should be near the rim and evenly spaced. Thread two strings through the holes (as in Make a Coat Hanger Balance - p.59). Tie the short strings to one long string and thread the long string through the pulley. Tie the pulley to a high place. Put a doll in the elevator and give him a ride.

How many ways can you find to use a rope and pulleys?

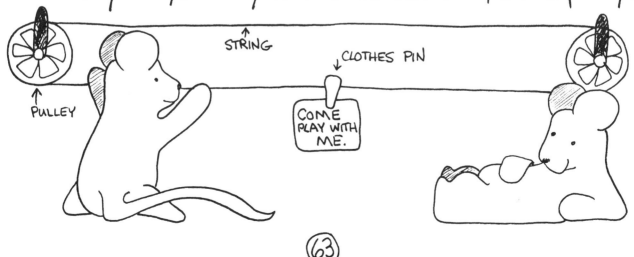

USE A GRINDER TO MAKE PEANUT BUTTER

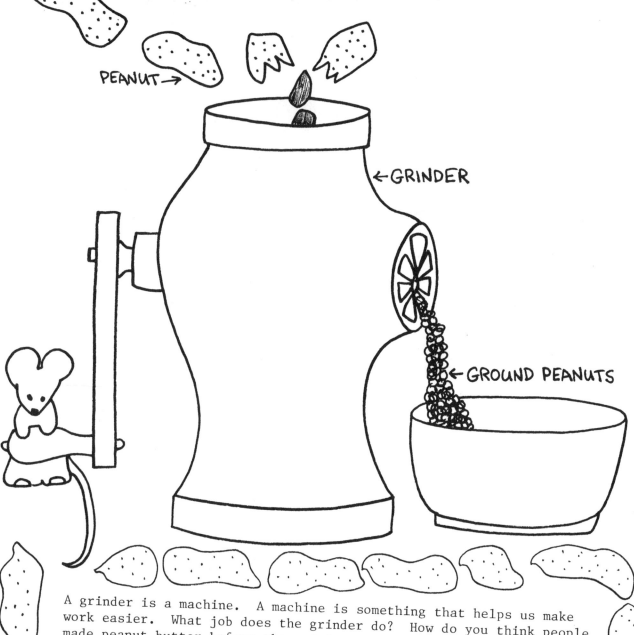

A grinder is a machine. A machine is something that helps us make work easier. What job does the grinder do? How do you think people made peanut butter before the grinder was invented?

Have you ever made ice cream with an ice cream maker? Noodles with a noodle maker? A milk shake with a blender? These are all machines that help in the kitchen.

Can you think of any other machines that you use at your house?

You will need:

A grinder - If you don't own one, try borrowing one from your neighbor.

Peanuts - Shell the peanuts yourself - it's fun!

Oil (peanut oil if available)

Honey

A bowl, spoon and knife

Put some nuts into the top of the grinder and turn the handle to grind them. Repeat until all the nuts are ground. Grind the nuts again to get them very fine.
   * VERY IMPORTANT - <u>Never</u> put your fingers into the top of the grinder! You might hurt yourself.

Mix the ground nuts with a spoonful of oil and a spoonful of honey. Taste and add more oil or honey as needed.

MAGNETS
PLAY A MAGNET SORTING GAME

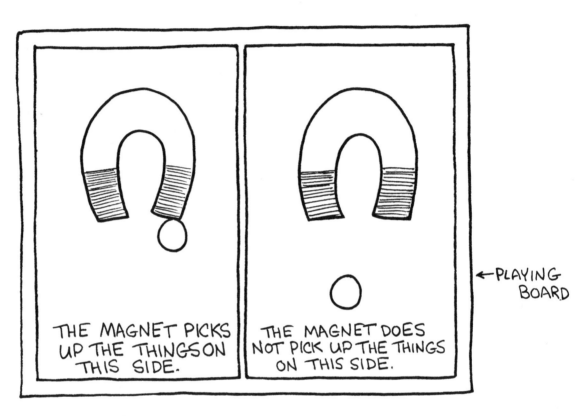

← PLAYING BOARD

THE MAGNET PICKS UP THE THINGS ON THIS SIDE.

THE MAGNET DOES NOT PICK UP THE THINGS ON THIS SIDE.

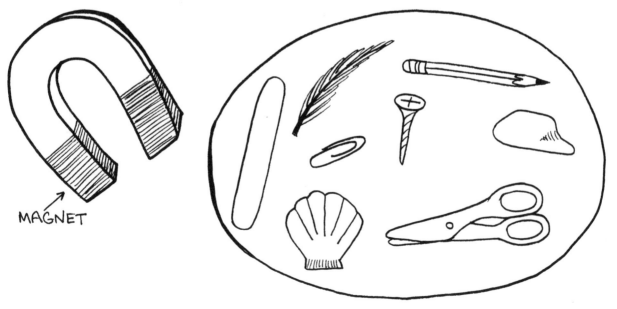

MAGNET

You will need:

A sheet of white paper

Felt-tipped pens or crayons

A magnet

HORSESHOE MAGNET   BAR MAGNET   CERAMIC MAGNETS

Assorted small objects made from a variety of materials.

Draw a playing board on the paper.

Select an object and guess whether or not the magnet will pick it up. Test the object and put it on the proper side of the playing board.

Repeat until all the objects are sorted.

Look at the objects which you have sorted. Can you explain why the magnet picked up certain things but not others?

Walk all around the room with your magnet. Try to guess if the magnet will attract a particular thing, and then test to find if you guessed correctly.

Keep a box of magnets and small metal objects such as screws and paperclips. By playing with your magnets you'll discover more about what they can do.

PLAY A MAGNET FISHING GAME

You will need:

- Colored construction paper
- Scissors, Paper clips
- Felt-tipped pens, A magnet
- A stick or dowel, String

Cut fish from different colored paper and draw on eyes and mouths with felt-tipped pens.

Attach a paper clip near the mouth of each fish.

Tie one end of the string to the dowel and the other to the magnet to make a fishing pole.

Scatter the fish on the floor and go fishing!

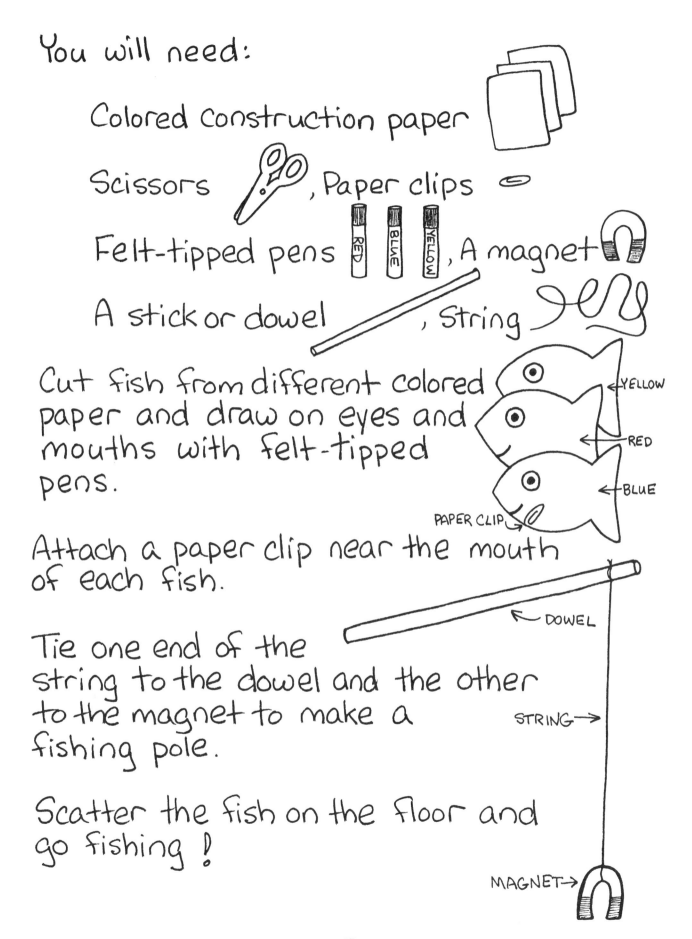

WATER SCIENCE IN THE BATHTUB

You will need:

Empty plastic bottles, paper cups, a funnel, an egg beater or anything else you can find around your house that looks as if it would be fun to play with in the tub.

*IMPORTANT — Never play with anything breakable in the tub.

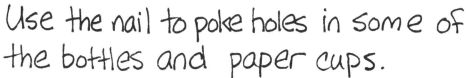

A nail, Food colors, Bubble bath soap

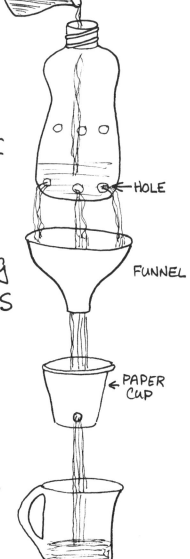

Use the nail to poke holes in some of the bottles and paper cups.

Add small amounts of food coloring to the water in some of the bottles and use the colored water in your play.

Put some bubble bath soap in the tub water and use an egg beater to beat up a big mound of bubbles.

SINK-FLOAT GAME

Look over the objects you have sorted. Can you explain why some of them sink and others float?

Float a toy boat. Drop small weights, such as pebbles or pennies, one by one into the boat. How many weights can you drop in before the boat sinks?

You will need:

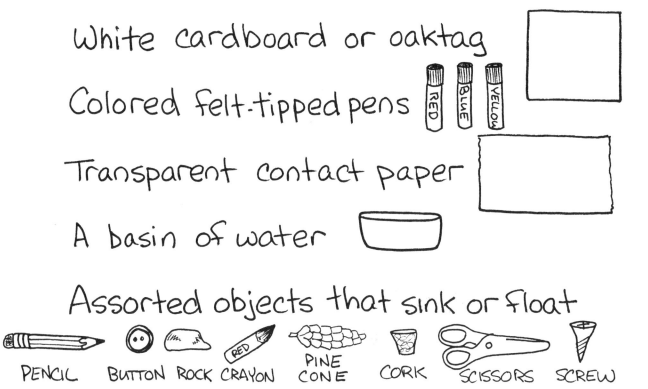

- White cardboard or oaktag
- Colored felt-tipped pens
- Transparent contact paper
- A basin of water
- Assorted objects that sink or float
  PENCIL BUTTON ROCK CRAYON PINE CONE CORK SCISSORS SCREW

Draw a sink-float chart on cardboard and cover the board with contact paper.

Guess if an object will sink or float. Test by putting the object in the water and put it on the correct side of the board.

Repeat until all the objects are sorted according to whether they sink or float.

POPSICLES

You will need:

Orange juice, grape juice or apple juice

An ice cube tray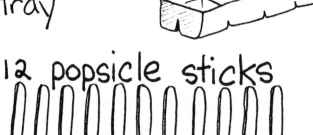

12 popsicle sticks

Aluminum foil

Fill ice cube tray with juice. Don't fill the tray all the way because the liquid will take up more room when it freezes.

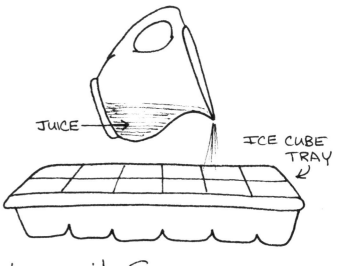

Cover the tray with a double layer of aluminum foil and crimp all around the edges to keep the foil in place. Stick a popsicle stick through the foil, into the center of each compartment. Carefully place the tray in the freezer.

When the popsicles have frozen, peel off the foil and carefully remove as many popsicles as you need. Leave the remaining popsicles in the tray and return them to the freezer for another day.

MAKE A BEACH SCULPTURE (WITH PLASTER OF PARIS)

You will need:

A small bag of plaster of Paris

A strip of cardboard (apx. 5" x 40")

Masking tape  A cup (for measuring)

A bucket , A spoon or stick for mixing

You can make a beach sculpture at the beach or at home in the sand box.

Bend the strip of cardboard into a circle or a rectangle. Overlap the two ends and tape them together.

Press the cardboard frame into the sand. Arrange shells, stones, driftwood, sea glass etc. on the sand inside the frame.

In the bucket mix 1 cup of water for every two cups of plaster. If you are at the beach, ocean water will work fine.

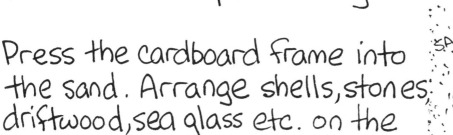

Mix the plaster with a spoon or stick until it is smooth.

Pour the plaster into the mold. The plaster should be at least 1½ inches thick. Mix extra plaster if you need more.

Let the plaster dry and harden. This will take about half an hour. Turn the sculpture over and remove the cardboard mold.

GERBILS

You will need:

A glass aquarium with a screen cover 10-15 gallon tank

A water bottle , A shallow dish

Cedar chips , Exercise wheel

Gerbil food , Cardboard tube

Two gerbils (Make sure they are the same sex - unless you plan to start raising gerbils)

Spread a one inch layer of cedar chips over the floor of the aquarium. Add the water bottle, food dish, exercise wheel and gerbils.

Give the gerbils tubes from rolls of paper towels or toilet paper or cut doors in empty cardboard boxes to make houses. The gerbils will chew up the cardboard for nesting material.

Gerbils are very clean animals, and you will not have to clean the cage more than once every few months. Allow the cedar chips and chewed cardboard to build up until there is enough for the gerbils to dig burrows.

Gerbils enjoy company, so it is a good idea to buy two; but remember, unless you want to raise gerbils you must make sure to buy two of the same sex.

In addition to regular gerbil food your gerbils love special treats such as peanuts, potato chips or small pieces of carrots and lettuce. They also enjoy getting out of their cage once in a while for some excersise. A good place to let them run is in the bathroom with the door shut.

The best way to pick up a gerbil is to grasp him by the base of the tail with one hand and lift him up on the palm of the other hand. If you always handle him gently, your gerbil will become a tame and friendly pet.

CRICKETS

You will need:

Crickets (available at pet stores that sell crickets as reptile food).

A clear plastic storage box with a lid or a glass aquarium

Two shallow dishes, saucers or jar lids

A sponge, A branch

A small piece of paper, Scissors

Crackers or bread crumbs, Water

Cut the sponge so that it fits into one of the dishes or lids. Moisten the sponge with water. The crickets will drink by sipping from the sponge.

Put a very small amount of crumbled cracker, cereal or bread in the other dish.

Put in a branch for the crickets to climb on and a small piece of folded paper for them to hide under.

Keep the box covered with a loosely fitting cover or a screen so that the crickets can't escape.

> Crickets shed their skins, so don't be surprised if one day you discover empty skins in the cage. Can you guess why this happens?
>
> If your crickets are happy in their new home they might start to "sing," which they do by rubbing their wings together.

ANTS

The easiest way to start a successful ant colony is to buy an ant farm kit at a pet store. The kit comes with a coupon which you mail in to receive your ants.

CATERPILLARS

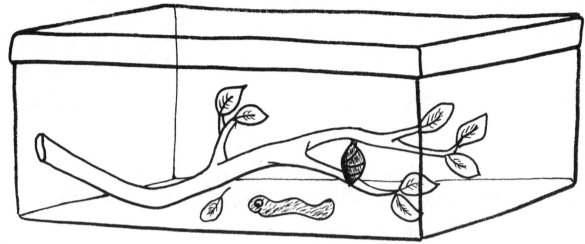

You will need:

A clear plastic storage box

A caterpillar

Leaves and small branches from the trees that grow near where you found the caterpillar.

—or—

A cocoon with the branch attached

- Put the caterpillar into the box with the branch and leaves. Feed the caterpillar by providing fresh leaves every day. After a while the caterpillar will build a cocoon around itself and change into a butterfly or a moth.

Do not disturb the cocoon. When the butterfly or moth emerges, take the box outside and remove the cover so that it can fly away.

Do you know how to tell the difference between a butterfly and a moth?

A butterfly rests with its wings folded and has little knobs on the end of its antennae.

A moth rests with its wings open and has no knobs on the end of its antennae.

WORMS

You will need:

A pail and shovel

A jar with a lid , A hammer and nail

The best time to dig for worms is after a rain shower when the soil is still moist. Dig up some dirt and go through it with your fingers. You might have to dig in several spots before you find any worms.

Poke several holes in the lid of the jar and fill it loosely with the dirt from the area where you found the worms. Put the worms in the jar. Keep the soil moist.